X-TREME FACTS: SCIENCE

THE OCEANS

by Catherine C. Finan

BEARPORT PUBLISHING

Minneapolis, Minnesota

Credits:

Cover, © Dmitry Brant/Creative Commons 4.0, © Sashkin/Shutterstock, and © Neil Bromhall/Shutterstock; title page, 17 top, cbpix/Shutterstock; 4, Guitar photographer/Shutterstock; 4 left, 4 right, Design Projects/Shutterstock; 4 center, Triff/Shutterstock; 5 top, Viacheslav Lopatin/Shutterstock.com; 5 bottom, 20 bottom, Joshua Avramson; 5 bottom left, Tomas Kotouc /Shutterstock; 5 bottom right, Sean Steininger/Shutterstock.com; 6 bottom, 8 bottom right, 8 bottom, 11 top, NOAA Okeanos Explorer Program, INDEX-SATAL 2010, NOAA/OER/Public Domain; 9 top, 13 top, 13 middle, 18 bottom, 19 right, 23 top left, 23 bottom, 24 bottom left, 26 bottom right, 27 top, 27 bottom left, NOAA/Public Domain; 7, Joshua Avramson; 7 top, Potapov Alexander/Shutterstock and pan demin/Shutterstock; 7 middle, superjoseph/Shutterstock; 7 bottom, NOAA/MBARI/Public Domain; 8 left, Gerringer M. E., Linley T. D., Jamieson A. J., Goetze E., Drazen J. C./Creative Commons; 8 top, Cromagnon/Shutterstock; 9 bottom left, 16 top left, NASA/Public Domain; 9 bottom, Richard Varcoe on behalf of Caladan Oceanic LLC/Creative Commons; 10 top, Berann, Heinrich C., Heezen, Bruce C., Tharp, Marie./Creative Commons; 10 bottom, gstraub/Shutterstock; 11 bottom, Lillac/Shutterstock; 11 bottom left, Narrissa Spies/Creative Commons; 11 bottom right, Laura Awasen Pasquinelli/Shutterstock; 12 top, Neil Bromhall/Shutterstock; 12 middle, Theodore W. Pietsch, University of Washington/Creative Commons; 12 bottom, David Csepp, NMFS/AKFSC/ABL/NOAA/Public Domain; 13 bottom, Olga Vasilyeva/Shutterstock; 13 bottom center, Lerner Vadim/Shutterstock; 13 bottom left, 13 bottom right, NORTHERN IMAGERY/Shutterstock; 14 top left, Esteban De Armas/Shutterstock; 14 top right, Paul Fleet/Shutterstock; 14–15 bottom, Wm. Leo Smith/U.S. Navy /Public Domain; 15 top, Toporkova/Shutterstock; 15 top right, Andrea Izzotti/Shutterstock; 16 top, Luka Peternel/Creative Commons; 16 bottom, Leslie Foot/Shutterstock; 16 bottom left, looisvee/Shutterstock; 16 bottom right, StudioSmart/Shutterstock; 17 bottom, chonlasub woravichan/Shutterstock; 17 bottom left, 17 bottom right, 26 bottom left, 28 bottom, Rich Carey/Shutterstock; 18 IFE, URI-IAO, UW, Lost City Science Party; NOAA/OAR/OER; The Lost City 2005 Expedition/Creative Commons; 19 top, 19 bottom NOAA Okeanos Explorer/Public Domain; 19 left, FlickDirect Inc/Shutterstock; 20 top, Igor Kruglikov/Shutterstock; 21 top, KGrif/Shutterstock; 21 middle, Mike Baird/Creative Commons; 21 bottom, Katharine Moore/Shutterstock; 22 bottom, Snowbelle/Shutterstock; 22 left, BeeRu/Shutterstock; 22 right, Krikkiat/Shutterstock; 23 top right, Kevin Raskoff/Public Domain; 23 bottom left, ©Citron/CC BY-SA 3.0; 23 bottom right, Bruce A.S.Henderson /Creative Commons; 24 top, J.S. Lamy/Shutterstock; 24 bottom, Kirk/Creative Commons; 25 top, scubadesign/Shutterstock; 25 middle, zaferkizilkaya/Shutterstock; 25 bottom, SHIN-db/Shutterstock; 25 bottom right, Roy Pedersen/Shutterstock; 25 bottom left, Artem Mazunov/Shutterstock; 26 bottom, Subphoto/Shutterstock; 27 middle, Eric Kilby/Creative Commons; 27 bottom, Melkov/Creative Commons; 27 bottom upper left, Vincent Lou/Creative Commons; 28 top, Ricardo Canino/Shutterstock; 28-29, Austen Photography

President: Jen Jenson
Director of Product Development: Spencer Brinker
Editor: Allison Juda

Produced for Bearport Publishing by BlueApple*Works* Inc.
Managing Editor for BlueApple*Works*: Melissa McClellan
Art Director: T. J. Choleva
Photo Research: Jane Reid

Library of Congress Cataloging-in-Publication Data

Names: Finan, Catherine C., 1972- author.
Title: The oceans / Catherine C. Finan.
Description: Minneapolis, Minnesota : Bearport Publishing Company, [2021] | Series: X-treme facts: science | Includes bibliographical references and index.
Identifiers: LCCN 2020039251 (print) | LCCN 2020039252 (ebook) | ISBN 9781647476755 (library binding) | ISBN 9781647476823 (paperback) | ISBN 9781647476892 (ebook)
Subjects: LCSH: Ocean--Juvenile literature. | Marine ecology--Juvenile literature. | Deep-sea ecology--Juvenile literature.
Classification: LCC QH541.5.S3 F49 2021 (print) | LCC QH541.5.S3 (ebook) | DDC 577.7--dc23
LC record available at https://lccn.loc.gov/2020039251
LC ebook record available at https://lccn.loc.gov/2020039252

For more information, write to Bearport Publishing, 5357 Penn Avenue South, Minneapolis, MN 55419.
Printed in the United States of America..

Contents

World of Water

Set sail around the world on a wet and wild adventure! With water covering about 70 percent of the planet, there's plenty to explore in the Atlantic, Pacific, Indian, Arctic, and Southern Oceans. On your journey, you can discover Earth's deepest point, its longest mountain range, and even an underwater forest. What else awaits? Maybe shipwrecks and sea monsters? It's time to lift anchor and sail on to find out!

The area of Earth's oceans is almost twice as large as all its land put together.

Earth's oceans hold over 97 percent of the planet's total water supply.

More of the moon and Mars have been explored than Earth's ocean floor!

The Arctic Ocean is frozen much of the year.

Oceanographers separate Earth's seawater into five oceans. But there is really just one huge ocean, and its waters are always moving around.

The Pacific Ocean covers about one-third of Earth's surface.

Earth's oceans are nature's highways. Humpback whales spend their summers in cold polar regions and swim to warm subtropical waters during the winter.

Get in the Zone

No matter which ocean you're sailing on, it has five zones. Think of it like a tasty layer cake! The part of the ocean we see—or the frosting—is the sunlight zone. Here, there is plenty of light in the water. Below this is the twilight zone, where things get darker! This is your first layer of cake. Go deeper still and you'll pass through more stacked layers in the midnight zone, the abyss, and the trenches. Let's go down, down, down—and explore!

Researchers use special deep-diving submarines to reach the bottom of Earth's oceans.

The sunlight zone

The sunlight zone is home to whales, dolphins, sharks, coral, and more.

650 ft (200 m)

The twilight zone

Sunlight barely reaches the top of the twilight zone, where jellyfish, squid, and octopuses live.

3,300 ft (1,000 m)

The midnight zone

WHO TURNED OUT ALL THE LIGHTS DOWN THERE?

Many creatures that live in the dark midnight zone, such as the deep-sea anglerfish, **can make their own light!**

13,000 ft (4,000 m)

The abyss

I DON'T KNOW, AND I DON'T CARE! WHO NEEDS LIGHT?

The near-freezing water in the abyss stretches to the seafloor.

Cusk eels live in waters up to 7,460 feet (8,370 m) deep.

The trenches

Seafloor trenches are the deepest, darkest places on Earth. Yet some life thrives there!

The Deepest Place on Earth

To go down deep, you need to set sail for the western Pacific Ocean. Below your boat (way, way below it) is the deepest point on Earth—the Mariana Trench! The deepest part of the trench, called Challenger Deep, reaches almost 7 miles (11 km) below Earth's surface. Just how deep is that? Earth's tallest point—Mount Everest—could fit inside it, and there would still be more than a mile of ocean on top of its peak. Now that's deep!

The **pressure** at the trench's bottom feels as though 50 jumbo jets were piled on top of you!

Some sea cucumbers and the Mariana snailfish can live in the Mariana Trench.

The Mariana Trench is 1,580 miles (2,550 km) long, but it has an average width of only 43 miles (69 km).

In 1960, the first people traveled to the depths of the Mariana Trench onboard a U.S. Navy deep-diving submarine.

Fewer people have made it to the deepest point on Earth than have walked on the moon!

Astronaut Kathy Sullivan is the only person who's **traveled to both space and the deepest point on Earth!**

Mountains Underwater?!

The ocean floor doesn't just dip down deep. It also has jagged points that shoot upward. In fact, Earth's longest mountain range lies beneath the waves! The mostly underwater mid-ocean ridge is longer than the Rockies or the Himalayas. The **tectonic plates** that make up Earth's crust are working overtime here, slipping and sliding on **magma**. As the plates move apart, **molten** rock bursts through the seafloor and builds up the mid-ocean ridge. Let's take an underwater mountain trek to explore!

The mid-ocean ridge's mountain chains stretch 40,390 miles (65,000 km) around the middle of Earth.

Mid-ocean mountains are steeper in places where magma moved more slowly.

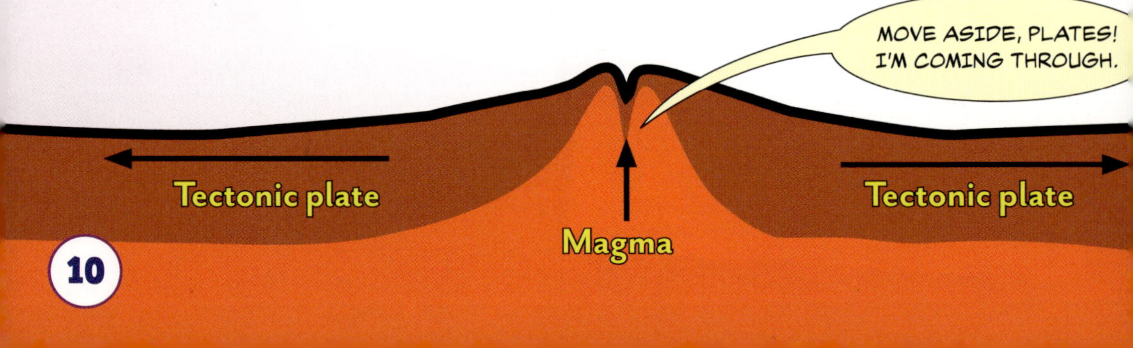

Most of Earth's volcanoes can be found along the mid-ocean ridge.

Once the scorching-hot magma reaches the sea floor, it cools off and turns into pillow-shaped lava.

Over 90 percent of the mid-ocean ridge is located in water so deep that sunlight doesn't reach it.

On average, it's over 1.5 miles (2.5 km) from the ocean's surface to the top of the mid-ocean ridge.

Fantastically Freaky Deep-Sea Creatures

We've found treacherous trenches and magnificent mountains in the deepest and darkest water. But what about the creatures moving around in those depths? Take sunlight out of the picture and things can get a little . . . freaky. Many deep divers make their own light—a talent called **bioluminescence**—to attract prey or escape **predators**. Others grow to gigantic sizes to help them survive the harsh conditions. Let's take a crazy, deep-sea creature tour!

The much smaller male anglerfish attaches to the female's body, using her blood and organs to survive.

Part of the female anglerfish's spine lights up and hangs over her hungry mouth, attracting curious prey!

The viperfish attracts prey with bioluminescence . . . and then **impales** it with fang-like teeth.

The blobfish's jellylike body is made for intense underwater pressure. Bring it to the surface and it just looks weird!

The giant Japanese spider crab can have a leg span of over 12 ft (3.7 m)!

Sea Monster Myths

Many of these strange deep-sea creatures could give you nightmares! But the stories of some underwater monsters are more **mythical** than real. Imagine being a sailor drifting on the open ocean with nothing but dark water all around. Who could blame you if your eyes start playing tricks on you? Throughout history, there have been tall tales about sea monsters that terrorize the oceans. Is there any truth to them? Where did these myths come from? And, most importantly, is it safe to go in the water?

The mythical kraken is a huge, many-armed sea monster that can drag ships underwater.

The kraken was probably based on the giant squid, a real sea creature.

What lonely and confused sailors thought were mermaids were probably manatees!

Stories of the monstrous leviathan describe it as a devilish whale. It is probably inspired by the gigantic sperm whale.

Many ancient maps feature scary-looking sea serpents. These were probably based on oarfish that can grow to 35 ft (10.5 m) long!

Crazy for Coral

If you escape the clutches of mythical sea monsters, sail on to Australia's northeastern coast. In the distance, the Great Barrier Reef stretches for nearly 1,400 miles (2,300 km) in the South Pacific. This is the world's largest coral reef system, and it's alive! But, no, it's not a mythical monster. Coral reefs are formed from living animals called polyps. Over time, the skeletons of large colonies of polyps join to form a reef.

The Great Barrier Reef is so large that it **can be seen from space!**

The Great Barrier Reef is the world's largest structure made by living things!

Green turtles, manta rays, octopuses, whales, sharks, and giant clams are all found in the **Great Barrier Reef.**

The reef is home to over 1,600 kinds of fish alone—like the cute clown fish!

Climate change threatens coral reefs. As water temperatures rise, delicate reef **ecosystems** can die.

Wet and Weird

You may visit the Great Barrier Reef and find amazing sea life. But just beyond it, much of the ocean is still uncharted. A recent research **expedition** to these waters discovered some amazing things, including entirely new animal **species**. The biggest—and strangest—find was a siphonophore (sih-FON-uh-fore), a ring-shaped, jellylike creature more than 150 ft (46 m) long. What other weird ocean life are we just finding?

The newly discovered siphonophore is about 50 ft (15 m) longer than the enormous blue whale!

Some deep-water siphonophores have orange or red digestive tracts that can be seen inside their see-through bodies.

Siphonophores use water jet propulsion to move around! Their bodies suck water in and squirt it out again to get them going.

The Dumbo octopus is named for its large, floppy fins that may remind you of the famous cartoon character's ears!

A 60,000-year-old cypress forest was recently discovered **buried in the Gulf of Mexico's seafloor.**

Does the Darth Vader sea cockroach remind you of a famous villain?

A newly discovered species of sea sponge looks like E.T., the famous movie alien!

A Forest under the Sea

You've come across canyons, mountains, reefs, and more on your underwater adventure. But those aren't the only things growing in the ocean. Just below the waves are huge forests—of kelp! Much like the Great Barrier Reef, a kelp forest is an important **habitat** for thousands of ocean animals, from crabs and fish to seals and sea otters. It gives many young ocean animals a safe place to grow up!

Seals and sea lions feed on squid, fish, and **mollusks** that live in kelp forests.

Giant kelp can grow over 175 ft (53 m) long!

Kelp is a kind of brown algae.

The ocean's algae, plants, and **bacteria** produce between 50 and 80 percent of Earth's oxygen! How can we ever thank them?!

Kelp's leaves grow balloon-like air bladders to help them float toward the surface.

Ocean animals eat kelp—and people do, too! **Kelp is used in ice cream and jelly.**

Sea otters wrap themselves in giant kelp to keep themselves in place while they sleep!

You're HOW Old?

There's plenty of new life in the sea. But what about the old? With its spiky shell and daggerlike tail, the horseshoe crab looks the same as its **ancestors** have looked for hundreds of millions of years! Horseshoe crabs and other creatures with ancient histories are sometimes called living fossils. While they aren't *really* fossils, they're almost unchanged today from what their relatives were like long, long ago—in some cases from even before the age of dinosaurs!

Horseshoe crabs have been around for 450 million years—that's more than 200 million years before dinosaurs walked on Earth!

The nautilus—an ocean mollusk related to the octopus—has barely changed in 500 million years!

Comb jellies have lived in the oceans for 700 million years— the longest of any animal!

Comb jellies are more than 95 percent water! They are basically just a blob of jelly surrounded by thin skin.

The frilled shark is a living fossil. The species has been around for at least 95 million years.

Deep-sea chimaera are closely related to sharks. They split off from their shark relatives around 400 million years ago!

Scientists thought coelacanths had died out with the dinosaurs 65 million years ago—until a fisherman caught one in 1938!

A Watery Grave

While some species have lived under the sea for hundreds of millions of years, the ocean has also been a place of sudden death. The power of stormy ocean waves can toss a huge ship around like a toy and send it and all its passengers to a watery grave. More than three million shipwrecks dot the ocean floor.

Coral and plants can grow on the remains of sunken ships.

The RMS *Titanic* sank in the North Atlantic in 1912 but wasn't discovered until 1985.

In 1986, a U.S. Navy deep-diving submarine called DSV *Alvin* carried people back to the *Titanic* for the first time since the famous ship sank.

The oldest intact shipwreck ever found was a Greek merchant ship that sank in the Black Sea around 400 BCE!

There are more historical artifacts in the ocean than are in all the world's museums put together!

In 1708, **the *San José*** sank off the coast of Colombia. **It was found in 2015—with billions of dollars' worth of treasure!**

There is about **$60 billion worth of sunken treasure** currently lost in the ocean!

Mysteries Beneath the Surface

Shipwrecks and sunken treasure aren't the only mysteries hiding beneath the waves. In fact, what we know about the oceans and the life they contain is . . . well . . . just a drop in the ocean! More than two-thirds of Earth's surface is under the water, yet 95 percent of it is still unexplored. You've learned a lot on your voyage, but so much remains to be explored and discovered!

More than 90 percent of known ocean species haven't even been classified yet.

Some scientists estimate that as much as **two-thirds of sea life hasn't been found yet!**

Frogfish

A new species of jellyfish

Recently, scientists almost missed the largest-ever underwater volcanic eruption. Who knows what else they haven't caught!

Little is known about the largest animal on Earth! **We're not even sure how long the blue whale lives.**

We don't know whether a strange pyramid and steplike structure off the coast of Japan is a human-made object or something from nature!

A bumpy purple blob recently found off the California coast might be a snail, slug, or jellyfish—or maybe a whole new species. Scientists don't know!

Sea Turtle Collage

Craft Project

There are over 300 species of turtles living on land and in water. But plastic pollution threatens the health of these animals. Make a sea turtle poster using the collage technique—and spread the message about plastic waste in the ocean!

Ocean animals may confuse plastic in the water for food they eat.

What You Will Need

- A sheet of white cardstock
- Watercolor paints
- A paintbrush
- Scissors
- A plastic bag (blue or clear)
- Craft glue
- A piece of paper
- A pencil
- Tissue paper
- A magazine or computer and printer

Step One

Paint a blue background on a sheet of cardstock using watercolor paints. Cut a plastic bag into strips. Brush glue on the painted cardstock and layer the strips over the paper. This will create a water effect. Set the background aside.

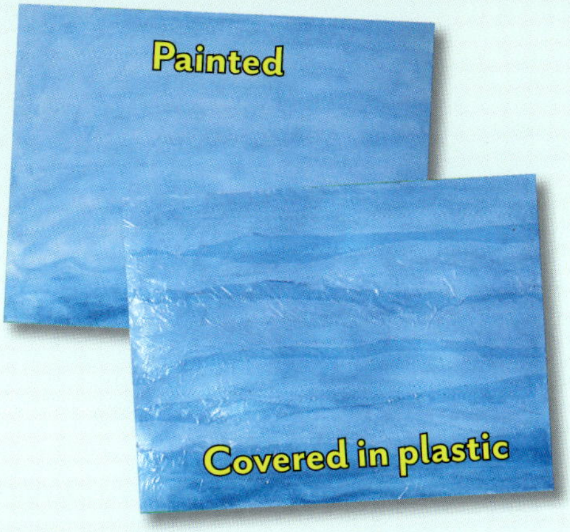

Step Two

Using a photo as a guide, draw a sea turtle on a piece of paper.

Step Three

Cut out the turtle shape. Scrunch up colorful tissue paper to create wrinkles. Glue the tissue paper on top of the drawing.

Step Four

Glue the turtle to the background. Cut out some letters from magazines or print them on a computer. Glue them to the background to spread your message about plastic waste in the oceans.

Glossary

algae plantlike living things that are found mainly in water and can make their own food

ancestors distant relatives

bacteria tiny one-celled organisms

bioluminescence light given off naturally by some living things

climate change the warming of Earth's air and oceans due to environmental changes, such as a buildup of greenhouse gases that trap the sun's heat in Earth's atmosphere

ecosystems the groups of living and nonliving things that make up an environment and affect one another

expedition a long trip taken for a specific reason, such as exploring

habitat the natural environment of an animal or plant

impales pierces with a sharp instrument

intact not broken or damaged

magma hot, liquid rock beneath Earth's surface

mollusks animals without a backbone that have a soft body usually enclosed by a shell; clams, shrimps, squids, and octopuses are mollusks

molten melted by heat

mythical based on a made-up story

oceanographers scientists who study oceans and the plants and animals that live in them

predators animals that hunt other animals for food

pressure the force made by pressing down on something, such as the force of water pressing from above

propulsion the force that moves something forward

species groups that animals and plants are divided into, according to similar characteristics

tectonic plates pieces of Earth's crust

Read More

Goldish, Meish. *Discovering Titanic's Remains (Titanica).* New York: Bearport Publishing, 2018.

Hutmacher, Kimberly M. *Lakes in the Ocean and Other Cool Underwater Facts (Mind-Blowing Science Facts).* North Mankato, MN: Capstone Press, 2019.

Mahoney, Emily. *20 Fun Facts About the Great Barrier Reef (Fun Fact File: World Wonders).* New York: Gareth Stevens Publishing, 2020.

Skerry, Brian, Elizabeth Carney, and Sarah Wassner Flynn. *The Ultimate Book of Sharks: Your Guide to These Fierce and Fantastic Fish (National Geographic Kids).* Washington, D.C.: National Geographic, 2018.

Learn More Online

1. Go to **www.factsurfer.com**
2. Enter "**X-treme Oceans**" into the search box.
3. Click on the cover of this book to see a list of websites.

Index

About the Author

Catherine C. Finan is a writer living in northeastern Pennsylvania. Her favorite ocean experience was scuba diving among coral reefs in the Bahamas. Her least favorite ocean experience was seeing a big barracuda up close and personal on that very same scuba dive!